Survival Writing
for Business

Steve Gladis, Ph.D.

HRD Press, Inc. • Amherst • Massachusetts

Published by: HRD Press, Inc.
           22 Amherst Road
           Amherst, MA 01002
           (800) 822-2801 (U.S. and Canada)
           (413) 253-3488
           (413) 253-3490 (Fax)
           http://www.hrdpress.com

ISBN: 0-87425-856-1

Cover design by Eileen Klockars
Production services by Anctil Virtual Office
Editorial services by Sally Farnham

DEDICATION

I dedicate this book to "Pop," Donald W. Sheehan, whose love of words and of his family were an inspiration to us all.

We all miss you, Pop.

TABLE OF CONTENTS

PREFACE

To write well, keep it clear and concise.
This book should help.

CHAPTER 1

USE DIRECT, USEFUL VERBS

In any sentence, verbs carry the action. They zap the sentence with electricity and set the pulse for written language. Because most people are visual learners, when you use action verbs, you'll help readers actually see that action.

Make Verbs Work

Verbs energize business writing, particularly verbs in their simplest forms: base verbs such as *hit, throw, drink,* and *stand* all convey the most direct meaning of the verb. When you convert the base verb to another part of speech, however, the strength fades and the color pales by comparison. So, let's say you're describing a company softball game:

Weakened form:	*Hitting* the ball over the fence, John scored the winning run.
Base verb:	John *hit* the ball over the fence and scored the winning run.

The first example just doesn't convey a strong visual message because the verb gets converted from an action word to a modifier (of John). Camouflaging the base verb reduces the action of the sentence. In the second example, you can hear the crack of the bat and watch the ball fly beyond the reach of the outfielders as it clears the fence.

Be Strong, Keep Order

A good simple sentence provides the cornerstone of solid business writing because it gives strength, clarity, and direction to language. Normal word order in a sentence is subject-verb-object: the verb cements the relationship between the subject and the object and grounds the three on a strong foundation. However, when that order changes, the relationships become less clear because the construction is weaker. In a sense, the foundation has cracked. Consider the following:

Out of order:	Authorization to pay the bill was given by the manager. (10 words)
Natural order:	The manager authorized the bill payment. (6 words)

The first sentence lacks direction and force because the order is inverted and the base verb is camouflaged as a noun.

Don't Hide Verbs

Camouflaged verbs are weak because they're abstract—you can't see, touch, taste, smell, or feel them. Consider two people who have just agreed to a contract after long and tedious hours of negotiation. Read the following two sentences and decide which one works better to convey meaning most directly:

Muffled negotiations:	After long hours of negotiations, labor leaders and management officials came to an agreement, and the contract was signed.
When two agree:	After long hours of negotiations, labor leaders and management officials shook hands and signed the contract.

The first sentence is difficult to envision. Most abstracts such as *agreement* are tough to visualize because they float somewhere beyond our senses. A handshake is a different matter, however. We shake hands every day to represent agreement. A handshake is concrete and familiar—an international sign that people agree.

Base verbs save money for the company because they are concise, direct, and easy to comprehend. Any time you can save paperwork and increase understanding, you put money into the bank. Look at these examples:

A cost overrun:	An illustration of the profit and loss statement was made by the accountant by the utilization of charts. (18 words)
An economical version:	The accountant used charts to illustrate the profit and loss statement. (11 words)

Besides being brief, the second example is easier to understand and is more to the point; as a result, the reader invests less and gets more in return—a perfect business deal.

To help you clear up clouded, weak writing, try this simple but effective exercise. Read a page of manuscript and underline every word that ends with -ent, -ant, -ion, -ment, -ence, -ance, -ency. Review the passage by deciding if the underlined words can be changed to the base form of the verb to improve the text. In most cases, you'll find that the meaning will become more direct and forceful. Try the underlining exercise on the following paragraph:

Bloated language: The company came to an <u>agreement</u> with labor to offer <u>employment</u> to minority groups across the board. The <u>decision</u> to offer jobs was reached on Sunday afternoon and the <u>implementation</u> was scheduled for the following Monday. The minority groups made a <u>statement</u> that, pending <u>implementation</u> of the new policy, they would withhold official comment. (55 words).

You should have underlined the following words: agreement, employment, decision, implementation, statement, and implementation.

Trimmed language: On Sunday, the company <u>agreed</u> to <u>employ</u> all minority groups. The minority groups, however, would not <u>comment</u> on the program before it <u>began</u> on Monday. (25 words)

The first paragraph contains 55 words and the second only 25 words—that's saving more than 50 percent. The first paragraph sounds fuzzy and vague; the second, clear and forceful.

The following is a short list of commonly camouflaged words and their base verbs:

Camouflaged Words	Base Verbs
Settlement	Settle
Assistance	Assist
Consideration	Consider
Arbitration	Arbitrate
Resistant	Resist
Negligent	Neglect
Performance	Perform

When you're trying to project the image of a company that deals straight and gets the job done without fanfare, which style do you think serves the company best?

Keep Your Voice Active

If real businesspeople should use base verbs, they should surely use the active voice as well. Why go halfway? To be direct and vigorous, use the active voice. Remember this cliché: Actives speak louder than passives. In the active voice, the subject does the action. In the passive voice, the subject is acted upon.

> **Passive:** The chef was fired by John. (6 words)
>
> **Active:** John fired the chef. (4 words)

In the second sentence John acts, but in the first the chef is acted upon. Not only is the second sentence more direct and consistent with the intended meaning, but it's also shorter. Brevity becomes a nontaxable fringe benefit of the active voice.

However, the passive voice provides good camouflage and a safe haven for company bureaucrats seeking ambiguity, anonymity, and obscurity. The writer who uses the passive voice avoids authorship and responsibility. Unfortunately, the passive voice has become a tradition in bureaucracies and has been passed down to each new generation like a curse. The tradition has become so entrenched that use of the active voice is often looked upon as odd. (Oops! I slipped into the passive voice, a sometimes legitimate practice if it serves a purpose such as varying the language. It's okay to use it for variety, but be careful that it doesn't become the norm.)

Indefinite approach:	The regulations were approved this past week and will be implemented immediately.
Definite style:	Jack Smith approved the regulations last week and will implement them immediately.

Reading the first sentence, the Chairman of the Board will not know who to call about the impact of the decision. The second sentence is not only more definite but also more personal. Save time and assign responsibility. Be direct and clear, and use the active voice—unless you're trying to cover up some obviously poor move. Then by all means use the passive voice, call in sick, or go on vacation.

Watch Out for *By's*

Passive language has crept so insidiously into our writing that it is often difficult to identify. To help you spot the obvious or hidden passives, use this aid: watch for all the *by's* in your sentence—even the implied *by's*.

Passive:	The redistricting of the sales territories was approved *(by someone)* and sent *(by someone)* to the president for approval. The plan was reviewed *by* the president's special assistant and finally approved *by* the president.
Active:	The sales board approved the redistricting of the sales territories and sent it to the president. His special assistant reviewed the plan before the president approved it.

The second paragraph tells the reader who approved what and gets through the matter quickly and with no confusion. The active voice is the standard rallying cry of any dynamic leader. If you use the active voice, your writing will clearly, directly, and forcefully represent you and your company. Don't "soft pedal" your wares. Ride to success by being more direct. Use the active voice.

Beware of Links that Leech

While most verbs convey action or state of being, some verbs merely link the subject to its modifier or state of being. These verbs, called linking verbs, are dangerous because they lack the vitality of active verbs. Read the following examples and see:

Drained:	John was happy.
Lively:	John laughed so hard that tears came to his eyes.

In the first example, *was happy* expresses joy, but only in a broad, bland way; it's unspecific and weak. On the other hand, *laughed* in the second example depicts something you can see and hear; it's more specific because it indicates the intensity of expression. Beware of linking verbs that steal the zest from your writing and sap its strength. Here is a list of links that leech:

- To be (is, are, was, etc.)
- Appear
- Become
- Seem
- Feel
- Grow
- Act
- Look
- Taste
- Smell
- Sound

In the business of writing, verbs are worth their weight in gold. Strong verbs give meaning and direction to the sentence and thus add value to your message. Cut liabilities, shore up values, and increase your profits: use base verbs, avoid the passive voice, and, where possible, shun linking verbs.

CHAPTER 2

KEEP SENTENCES SHORT

When is writing like preparing food? Some would say, "When you eat your words." Actually, writing is a process like food preparation; both take time and cost money. Consider the following scenario: your boss has just told you to answer a letter or an e-mail that she received from the ABC Company complaining and seeking a refund for one of your company's defective widgets. Look now at how the process of writing a response costs your company money.

First, you spend 10 to 15 minutes to decipher the boss's note on the bottom of the page. Finally, you relent and go to the source. After a long-winded sermon on what it means to be a self-starter, she translates her comment: "Handle this." Bewildered, you return to your office.

So, you reread the letter, and wonder if the boss really has any idea of either the problem or what she wants done. After a trip to the company library for background information, you lunch with a buddy and relate your stimulating conversation with your boss. Your friend offers to give you a copy of a similar letter he wrote a week ago. Though the topic differs, the format helps you write an initial draft. After revisions and a few tips from your buddy, you finish the letter and send it along to your boss. She sends it back to you with a few additions and deletions of her own. Eventually, you send the letter.

The Economic Impact of Writing

If you determined the total processing cost of your letter, you'd be surprised. Some estimates range conservatively at $20 per page for typical business communications. Thus, when you write, you want the process to be quick and efficient and the final product to be accurate, succinct, and understandable. If not, your company pays dearly.

If you multiply the millions of e-mails, letters, and reports written every day in government and business by $20, you can see the fortune to be saved by trimming back the flow of unnecessary words. Hence, the writing process does have an economic impact.

The Rule of Thirty

Tackling business paperwork can intimidate the average person, even when all your biorhythms are positive. So, start small: begin with the basic sentence. To contain the verbal explosion in business, apply the Rule of Thirty to your sentences: Keep sentence length to thirty words—or no more than three typed lines—and you'll hold your reader's interest. The Rule of Thirty necessarily trims back the fat and reveals your message. Readers cannot easily comprehend inflated, tedious sentences. They lose interest—fast. Consider the following example:

Word proliferation:	This letter is written concerning our product that was the subject of your most recent inquiry concerning our manufacturer's rebate which we do give to such well-regarded customers like your company which has enjoyed a long-standing relationship with our firm. (44 words)

This sentence would challenge even the most interested reader. Long, overwritten, and boring, it loses its thrust because of its wordiness. To help you cast off its excess baggage, use the Rule of Thirty. Begin by underlining all the key words—those words essential to your message. In this sentence, you would probably underline the following words: *product, manufacturer's rebate, well-regarded customers,* and *our firm.* Next, write the sentence in more direct language, still keeping the message that good companies get rebates.

Economical version:	Yes, our firm will offer you a product rebate because you are one of our preferred customers. (16 words)

The new shortened sentence is only sixteen words long. We have saved twenty-eight words. That is a cost reduction in anyone's book. The Rule of Thirty, however, is not chiseled in granite and should be used only as a guideline and not as a commandment. Remember when trimming back excess, underline key words, choose language that is active and expressive, and keep your sentences to no more than thirty words.

The Way to Be Brief

To help you trim back excess words, you should:

Cut Prepositional Phrases

These causes of grammatical obesity are as fattening as chocolate-covered strawberries. For example:

Obese	Trim
Along the lines of	Like
As of this date	Today
√ At the present time	Today

(continued)

Obese	Trim
For the purpose of	Because
Inasmuch, as	For, as
In order to	To
In relation to	About
In the event that	If
On the basis of	By
On a few occasions	Occasionally
Subsequent to	After
With reference to	Concerning

To cut back wordy prepositional phrases, underline all prepositions such as by, with, at, in, to, on, for, between, and from. Then, sharpen your editorial carving knife and slice the fat away from the lean.

Delete Double-Dealing Words

Wasted time costs money in any business. Duplication of effort, whether on the job or in writing, wastes time for everyone concerned. *Tautologies* are clusters with overlapping meanings. They duplicate efforts, bloat language, and waste time. The following are examples of tautologies and efficient solutions to them:

Tautologies	Solutions
Advance planning	Planning
Ask the question	Ask
Continue on	Continue
Cooperate with	Cooperate
Protest against	Protest
Definite decision	Decision
Free gift	Gift
Free pass	Pass
Invited guests	Guests
Old adage	Adage
Personal friend	Friend
Dash quickly	Dash
True facts	Facts
Reason why	Reason

To ferret out tautologies, look at the following example for nonessential words, and you will find at least part of the tautology:

Tautology trial: During the anti-trust trial, the president gave the true facts about the conglomerate's merger plan.

Note that *fact* is a word essential to the sentence, but *true* is not. You are more likely to find tautologies lurking among the unnecessary words and often right next to an essential word. With a sharp eye, you can spot these duplicators, then flush them out of the sentence.

Beware of the Passive Voice

As we discussed in the previous chapter, use of the passive voice not only will sap writing of its vitality, but will also bloat your writing. The passive voice construction adds extra words—verbs of being (is, was, were) as well as qualifiers—and it often leaves the reader with questions of authorship.

Original bid: It is recommended that the company accept the bid submitted by the contractor.
(13 words)

"Who recommends?" is the first problem that the sentence poses for the reader. The second problem: The writer must explain which bid to accept and must use extra language to do so.

> **Final bid:**   I recommend that we accept the contractor's bid.
> (8 words)

This sentence neither generates questions nor requires explanations; moreover, the sentence is more than 30 percent shorter. Increased clarity and word economy deposit money in the company's bank account.

Economic inflation can be compared to writing. Just as prices increase, eroding our buying power, so do words as they multiply, eroding the power of our writing. When you see word clusters creeping into your text, stop the pile up by leaving them out.

CHAPTER 3

AVOID ABSTRACTIONS AND USE CONCRETE LANGUAGE

Most of us remember the old Bible tale of how God gave Moses the Ten Commandments on a mountaintop. As the story goes, God etched the Commandments on two large stone tablets, and they became the best known laws in history. Why? Because they are direct, concrete, and to the point, and they make sense. Just consider one of the Commandments as an example: *Thou shalt not kill.* Though the language is dated, it leaves no doubt in the reader's mind about what is expected.

On the other hand, what if the Commandment read like this:

> Having been a party of or an accomplice to the willful taking of human life, whether patricide, fratricide, or homicide of any type, and whether the deed be premeditated or spontaneous, such act presents the gravest of moral, ethical, and spiritual ramifications for the malefactor.

Such bloated language can have an impact. First, Moses would have needed a moving van to haul the tablets from the mountaintop. Second, no one would have ever understood the Ten Commandments.

Abstractions usually confuse readers. This is not to say that abstractions have no place in our writing. To the contrary, abstractions represent man's ability to deal with pure ideas, to universalize, and to do higher-order thinking. However, except for philosophers, words lose clarity as they lose their color and sense.

Tangible words come to us through the senses—we hear them, see them, reach out and touch them. As we move away from the sensual experiences and toward intellectual ones, however, we create abstractions. Because each of us forms abstractions differently, diversity and confusion often result. By moving out and away from the senses, writing often loses its focus, leaving the reader behind in the process.

The Ripple Effect of Abstractions

Abstractions are like the concentric ripples that form when you throw a rock into a pond. A rock entering the water is like the solid, concrete image. Once the rock breaks the surface, shock waves are sent out in concentric ripples, each bigger than its predecessor, yet less defined. To illustrate the "ripple effect" in language, let's consider something critical to business: money.

To begin with the most concrete image of money— analogous to the rock splashing into the water—picture a new one dollar bill. You can see it and feel it. You can fold it, put it in your wallet, and buy a candy bar with it. It's real. But as the ripples move out from the 6" × 2½", green one dollar bill, the images get softer, more unfocused, and more difficult to comprehend. Look at the following list to see the ripple effect on money:

1. dollar bill
2. folding cash
3. money
4. funds
5. currency
6. medium of exchange

You can see that as language moves away from the concrete dollar bill, it devalues rapidly, and its meaning gets obscured in the process.

Avoid unnecessary abstractions in your writing by employing the Keep-It-Simple rule. When in doubt about using abstractions, journalists ask these questions: Who, What, When, Where, Why, and How? Answer these questions about each piece you write, and your writing will bristle with concrete language. Look at the following sentences:

Devalued language: The business executive traveled out of town to attend a meeting last week.

In this example, abstractions and generalities are overused. *Executive, traveled, out of town, meeting,* and *last week* are all abstract words that tell only a vague story. Because of abstractions, the reader naturally asks: Who went out of town? How did that person travel? Where and why did that person leave? The questions raised by abstractions bog down the reader and waste too much time on images that should be in clear focus from the start.

Value-based language: John Sloane flew to Denver last Thursday to attend the annual stockholder's meeting.

In the second example, the questions raised by the first are answered merely by using direct, concrete language. Each critical word becomes like a rock splashing into the water and breaking its surface with crisp ripples. This sentence tells the readers everything they want to know.

More Cloudy Language

Excessive variation also clouds up business writing. To understand this term, you must think back to elementary school days when a teacher warned you not to use the same word twice in the same sentence. That instruction focused on improving students' vocabularies. But in an adult world, constant variation (excessive variation) often leads to confusion. Consider the following sentences:

Variation on a theme:	One executive in the company received a handsome bonus, while another supervisor got a lesser amount, and yet another manager got even less than the others. (26 words)
Cleared up:	The three company managers received three different bonus amounts. (9 words)

In the first example, *executive*, *manager*, and *supervisor* are intended to mean the same thing—but the reader is not really sure. The first sentence avoids repeating the same word three times, but confuses the reader in the process (excessive variation). The reader will naturally wonder: Are these three company officials of the same executive level? Thus, the reader has to weigh each variation against the others and in many cases is driven to an unabridged dictionary for definition of esoteric words. Simple message: Don't show off your wide vocabulary at the reader's expense.

The Value of Concrete Language

People read to learn, not to be impressed, insulted, or even belittled. Overwriting—a close cousin of excessive variation—can also inhibit meaning. Of course, I am not suggesting that you write only in monosyllables. Some language variation adds spice to writing and makes the text more palatable. Just remember that too spicy a dish causes indigestion. Read the following examples:

Stuffy stuff: Due largely to a paucity of concern, this day has been summarily terminated.

Aired out: Because of a lack of interest, today has been canceled.

These two versions need no explanation. We have all been guilty of showing off. Avoid confusion for the reader by using clear, familiar language to convey a specific message.

Good business operates best in a concrete world. To be effective in business, you must communicate your company's ideas in words that people understand quickly and clearly. The surest way to relate to others is through their senses—words that help readers touch, taste, smell, see, and hear. Use concrete and familiar language; strive to communicate, not impress.

CHAPTER 4

KEEP RELATED
WORDS TOGETHER

Natural order rules the business world: expenses usually come before profits, design before production, and marketing before sales. Language also has an order. Standard word order in the English language is subject-verb-object.

Standard sentence: The banker (subject) gave (verb) the money (object) to him.

In fact, English relies on word order for meaning. Consider the following sentences if you have any doubt:

Bow: The dog bit the man.

Wow: The man bit the dog.

Important Words Go First or Last

"Put your best foot forward." Here's a saying that provides wise advice for life and writing. People remember what they hear first—the law of primacy (first). So place important words at the beginning of a sentence.

Buried effort: This year we *reduced expenses* by monitoring our spending habits.

Up front: *Reduced expenses* resulted from monitoring our spending.

"Save the best for last." Here's another saying. This one demonstrates the rule of recency—say important words last to emphasize them to readers. Suppose you wanted to emphasize to your boss that certain efforts have resulted in big profits for the company. The key word is *profits*. Consider the impact of the next two sentences:

Muffled profits:	In the recent past, our efforts to increase *profits* by cutting overhead expenses have been successful.
Roaring profits:	Recently, our successful efforts to reduce overhead expenses have increased *profits*.

In the first example, *profits* gets buried in the sentence, and the reader must either reread it to get your point or lose it among the words. The second sentence, however, makes sure that the reader gets the point by placing *profits* in an obvious and prominent position—last.

To sum up: If you want to emphasize a word, either put it at the beginning or at the end of a sentence. If you choose a place other than these two locations, your emphasis will be flattened, perhaps lost. Just remember another old saying: "Position is everything in life!"

Misplaced Modifiers

Chalk up one of the most common errors in business writing to misplaced modifiers. Adjectives, adverbs, phrases, and clauses all modify or describe something in a sentence. When misplaced, they modify the wrong thing and mislead the reader:

Misplaced modifier:	*Typing the report furiously,* the boss interrupted *me.*
Realigned order:	*While typing the report,* I was interrupted by the boss.

The first sentence has the wrong person doing the typing. By moving the typing closer to the real worker bee, *you* get the job done faster.

As a rule, keep the modifiers and related words together in a sentence to avoid confusion like this:

Poor management:	The CEO discussed filling the vacancies in the mail room with the senior partners.

In this sentence, it looks as if the boss is going to place some pretty high-priced talent in entry-level positions. What the writer probably intended to say was:

Unruffled partners:	The CEO and the senior partners discussed the vacancies in the mail room.

By moving the senior partners closer to the boss, you will save yourself an explanation memo to some ruffled executives.

Misplaced Words

The simple way to solve ambiguity that arises from misplaced words is to underline the modifiers in question. Underlining helps you see the distance between the modifier and the word being modified. Gaps between the two will show up immediately.

Misplaced salesman:	The <u>salesman</u> made an important sale to a major client <u>using visual aids</u>.

In this sentence, the client seems to be using the visual aids instead of the salesman, which is pretty good salesmanship if you can do it! The following sentence, however, is probably more accurate:

Good salesmanship: Using visual aids, the salesman made an important sale to a major client.

Squinters

When modifiers can modify words on either side of themselves, they are said to "squint," because they do not focus well on the word you want to modify. When in doubt, place the modifier nearest the word you want to modify—and as far away as possible from any other word with which it could be mistakenly linked:

Squinting: The committee that was studying redistricting of sales territories *yesterday* made its recommendations to the national sales manager.

Open eyed: After studying redistricting of sales territories, the committee recommended changes *yesterday* to the national sales manager.

After reading the first example, you might think that such a major reorganization was done rather haphazardly and in a rush because *yesterday* looks like it refers to how long the committee met—*one day*. Yesterday, however, refers to when the final report was made and not to the duration or time spent studying. That is a big difference! If you were the boss, wouldn't you have some questions about the quality of the report after reading the first sentence?

Phrases and Clauses

Dangling phrases or clauses hang freely without being clearly connected to whatever they are intended to modify. Such misplaced modifiers can lead to unwanted humor and confusion among readers:

Nervous boss:	After straightening his tie, combing his hair, and clearing his throat, the *boss* was briefed by the newest partner in the firm.
New partner:	After straightening his tie, combing his hair, and clearing his throat, the *newest partner* briefed the boss.

In the first sentence, the boss appeared to be going to a lot of trouble to impress the new partner. The second sentence makes more sense. When a phrase or clause comes at the beginning of a sentence, the word modified should come after the comma or as close to it as possible to avoid confusion.

Sentences

A gap in a White House tape caused a certain president some real problems, and a gap in your taxable income causes you big problems. Remember one simple rule to avoid confusion in writing: gaps cause trouble. Keep subjects and objects close to their verbs to avoid gaps and confusion.

The gap: The free *trip* to Hawaii, which was the grand prize given each year by the TDY Company for the salesperson who sold the most widgets to the federal government, *was given* to Samantha Jones.

The reader gets lost en-route while trying to get the trip to Hawaii and Samantha Jones together because the subject, *trip*, and the verb, *was given*, are separated by a wide gap. Try this sentence as a solution:

Closing the gap:	The *trip* to Hawaii *was given* to Samantha Jones as the grand prize for selling the most widgets for the TDY Company to the federal government.

This sentence helps bridge the gap to Hawaii by getting the subject, *trip,* and verb, *was given,* together. This way, Samantha gets her deserved reward more quickly and clearly. Note that the sentence could also be rewritten with Samantha Jones as the subject if she were to be emphasized and not the trip.

Another way:	Samantha Jones won the Hawaii trip as the grand prize for selling the most widgets for the TDY Company to the federal government.

Consider the following sentence and try to find the gap between the main verb and its verb phrase:

> **Gap:** The managing editor tried, for literally years while employed at the newspaper in the editorial group, to be objective and fair.

By splitting the verb *tried to be*, the writer creates a gap in this sentence that may lead the reader to think the entire editorial group is trying to be objective and fair—not a bad idea but not the intent of this writer. Simply put: The reader has to vault over too much territory to land safely on the other side of this gap. Shorten the verbal gap like this:

> **Bridged:** While employed for years in the newspaper's editorial group, the *managing editor tried to be* objective and fair.

Now there is little question who is trying to be objective and fair. Sort out the verb and its object in the example below if you can:

> **Hedging:** The Chairman of the Board announced, after bragging about the wonderful hotel accommodations, the fabulous social galas for the week, and the golf and tennis arrangements, a deficit for the year.

This sentence makes it appear like the golf and tennis arrangements had a deficit and not our unfortunate company. Try this revision:

> **Facing the music:** After bragging about the wonderful hotel accommodations, the fabulous galas for the week, and the golf and tennis arrangements, the Chairman of the Board *announced a deficit* for the year.

If you are trying to hide the deficit from the stockholders, by all means use the first sentence! If, however, you want to make sense, be honest, and not confuse your reader, use the second sentence.

So to avoid confusion and to increase clarity:

- Keep related words together.

- Stress words by placing them at the beginning or end of the sentence.

- Keep modifiers close to the words they modify.

- Underline to find your word gaps.

- Do not let modifiers dangle.

- Keep subjects and objects close to their verbs.

CHAPTER 5

DON'T SHIFT NUMBER, TENSE, VOICE, SUBJECT, OR POINT OF VIEW

Shifting galloping horses midstream may be all right for cowboys, but do not try it when writing, or you may get all wet. Any time you shift number, tense, voice, subject, or point of view in a sentence, you stand a good chance of dumping your readers as well.

Shifting Numbers

Pronouns refer back to nouns. He, she, it, we, they, and other pronouns substitute for nouns and lend variety to your writing by avoiding the needless repetition of nouns. When used incorrectly, however, a pronoun can stop readers in their tracks.

Dwindling investors:	The *investors* were interested in buying the property until *she* saw the city dump adjacent to it.
Consistent investors:	The *investors* were interested in buying the property until *they* saw the city dump adjacent to it.

In the first example, the *investors* shifted to *she*, and suddenly a conglomerate became an individual. The reader gets confused and certainly loses confidence in a group of investors that dwindles so quickly to a single *she*.

The second example starts with a group, *investors,* and ends with a group, *they.* If you had to invest in either of these two sentences, where would you put your money? Make sure that pronouns agree with the words they relate to. Leave the shifting to your sports car, not your writing.

Indefinite Pronouns and Language

An indefinite pronoun does *not* refer to a specific person or thing. Pronouns such as each, either, everyone, everybody, somebody, and nobody are only some examples of indefinite pronouns. All are singular.

Because they are frequently mistaken for the plural, they are often the sources of unending errors in writing. Consider the following examples:

Definitely confusing:	Everyone brought *their* coats to the meeting.
Clear cut:	*Everyone* brought *his* or *her* coat to the meeting.

The first example creates a confusing image for the reader. No one is sure whether each person has several coats or the group owns a few coats. The second sentence makes it clear that each person brought one coat—*his or her coat.* Indefinite pronouns do not have to cause indefinite language if you stay on track and use the singular. If you want to avoid the his/her construction, turn the sentence completely plural.

An alternative: All brought their coats to the meeting.

Here, you retain an accurate accounting of the entire group.

Collective Nouns

You remember the Class of 2000 and the Olympic Team? How about the Senate and House of Representatives? The committee, the group, the council, or the class—they are all collective nouns. When you use them in a singular way, use a singular pronoun; use a plural pronoun if you are referring to each individual in the group. Just be consistent.

All for one: The *committee* on waste *works* well, and *it saves* the company inestimable amounts of money.

One for all: The *committee vote their* consciences on all issues and *are* very independent.

In the first example, *committee* is used collectively and, therefore, takes the singular pronoun *it* as well as the singular verbs *works* and *saves*. The second example uses *committee* to refer to the individual members of the committee; hence, committee is used in a plural sense and requires the plural pronoun *their* and the plural verbs *vote* and *are.*

Shifting Tenses

Time means money in business. Deadlines rule industry, and if you miss them constantly, you might consider the advantages of unemployment. For executives, time is especially precious, and a host of books and courses teach leaders how to save it.

When you shift tenses, you force readers to reread, reflect, or refigure in order to understand your message. If readers refuse to spend such time, they never get your message. Time is limited, so don't waste your readers'. Pick a tense and stick to it.

Timing problem:	The CEO *studied* the proposal thoroughly and *makes* a wise decision.
In sync:	The CEO *studied* the proposal thoroughly and *made* a wise decision.

The first example shifts tenses from the past tense, *studied,* to the present tense, *makes,* leaving the reader caught in a time warp and uncertain about when the action took place. The second keeps all the action in the past tense, *studied* and *made,* thus, saving readers the effort of time travel.

Changing Voices

Shifting from the active voice to the passive voice also confuses most readers and appears as immature and as laughable as the squeak of an adolescent's voice.

Voice change:	The committee chairperson *opened* the meeting after the notes *had been reviewed.*
Straight talk:	The committee chairperson *opened* the meeting after *she* reviewed the notes.
Even straighter talk:	After *reviewing* the notes, the committee chairperson *opened* the meeting.

The first sentence causes the reader to wonder who reviewed the notes. The second sentence is clearer and leaves little doubt: The reader is able to follow the sentence and certainly gains confidence in the chairperson. The third sentence cuts even closer to the bone. Placing the phrase up front and next to the chairperson leaves no doubt about who *reviewed* the notes and then *opened* the meeting. Speak loud and clear and speak with one voice.

Changing the Subject

Sometimes in business, changing the subject or topic serves as a useful tactic. It will definitely throw people off course, divert attention, and shift the emphasis to less dangerous areas—perhaps the safest course, especially in sticky situations. However, if you want to be readily understood by others, stick to the subject in all of your sentences:

Subject shift:	*Mr. Edwards* corrected the confusing report, and the *report* was retyped.
Consistent:	*Mr. Edwards* corrected the confusing report, and then *he* had it retyped.

In the first example, the subject shifts from *Mr. Edwards* to *report*. The reader wonders if Mr. Edwards or someone else typed the report. You should also note that a confusing shift in subject often accompanies the passive voice, as is true in this example. The second sentence smoothes out the bumps and keeps the subject consistent throughout the sentence: *Mr. Edwards . . .* and then *he . . .*

A Matter of Point of View

You may remember that point of view refers to the perspective from which a story is told. For example, if a narrator is one of the main actors in a story, she would use the first person, *I* or *we*, as the point of view. If another narrator directly addresses the reader, he would use the second person and say *you*.

Finally, if narrators tell stories about others performing the action, they write about *he, she, it,* or *they.* Writers will often shift the point of view in mid-sentence and leave the reader wondering who's telling the story.

Point shift:	The *boss* said that when *I* read the report of the auditors, *you* could see the obvious errors that *we* had been making.
Right on point:	The *boss* said that when *she* read the auditor's report *she* could see the obvious errors that the company had been making.

The first example shifts from the *boss* to *I* to *you* and back to *we,* and the reader falls between the cracks, unable to decide who is in charge of this sentence. In the second example, it is clear that the *boss* is in charge. The point of view clearly is her own.

Shifts in point of view are common, particularly in longer pieces of writing. These shifts usually occur when the writer loses sight of who's telling the story and the intended effect. Keep your writing on track through proofreading and by asking yourself: Who is telling the story?

Gender-Biased Language

Words like *businessman, chairman, workman,* and others like them creep into our writing and reflect a male bias, particularly in the business world. These gender-biased words presume that men dominate the workplace. Gender-biased language denies women their rightful status, and it should be scrupulously avoided. Here are some tips that will help:

Avoid Gender-Dominated Idioms

- a man-sized job
- a real he-man
- a weak sister
- the manly thing to do
- a woman's touch

Use Plural Subjects to Avoid Gender-Biased Pronouns

Not: Each *editor* took his reporters to meet the publisher for the first time.

But: All the *editors* took their reporters to meet the publisher for the first time.

Avoid Male-Dominated Titles

Not	But
Policeman	Police officer
Chairman	Chairperson √
Workman	Worker

Use Job Titles to Avoid Servile Gender Roles

Not: The *woman* behind the counter offered *her* help by giving directions to the customer.

But: The *attendant* helped by giving directions to the customer.

Omit the Pronoun Where Possible

Not: If treated well by management, the American worker does *his* best on the job.

But: If treated well by management, the American worker does very well on the job.

Use *He and She* or *He/She* if You Must

Use this technique only as a last resort because it is awkward and contrived. However, it's far better than a gender-biased alternative.

Not:	Each worker turned in *his* hours to the timekeeper.
But:	Each worker turned in *his or her* hours to the timekeeper.
Better yet:	The workers turned in their hours to the timekeeper.

Last Word on Shifts

There are a lot of shifts in life. There are shifts in the tide and shifts in the land. We work different shifts and often shift positions to get comfortable during a boring meeting. We know shifty businesspeople and try to avoid them. We experience violent shifts in the stock market and try to escape with our skin.

There are shifts in football, cars, trains, planes, and buses. We shift, they shift, and you probably shift, too. But every time you see a shift, there is great potential for a slip. When you feel a shift coming on in your writing, remember: The slip you save may be your own.

Stamp out shifts in our time: Avoid shifts in number, tense, voice, subject, and point of view—and avoid using shiftless gender-biased language.

CHAPTER 6

STRUCTURE YOUR WRITING

If you have ever taken the time to review your company's organization, you have probably noticed several things about its structure. First, you may have observed the complexity of the corporation. Even the structure of mid-sized U.S. corporations can be remarkably intricate and specialized. Second, you may have noticed that the hierarchical order of the structure makes it easy for you to determine both the corporate power structure and personnel functions. And, third, you probably realized how far most of us have to go to make it to the top.

Parallelism

Any good corporate organization chart is built on a simple, effective foundation: parallelism. Department heads stand next to other department heads and vice presidents stand with other VPs. In writing, the basic principle of parallelism is this: Similar and logically related ideas must be expressed in the same grammatical structure. Parallelism pumps vitality into any good piece of writing because it leads the reader directly from likeness in form to the likeness in both content and function. Consider how quickly you can tell who's who in your corporation just by their positions in the organizational chart.

Parallelism makes writing more effective and efficient because it:

- Forces the writer to think logically
- Gives structure to writing
- Makes sense to readers

Most importantly, parallelism makes writing easier to understand.

Unparallel lines:	The line supervisors' responsibilities are training of new salespeople, to develop new customers, and support upper-level management's decisions.
Parallel lines:	The line supervisor's responsibilities are to train new salespeople, to develop new customers, and to support upper-level management's decisions.

In the first example, the grammatical forms of the three responsibilities are different and confusing: *training, to develop,* and (to) *support.* Note that all are verb forms that act as nouns, but because they look and sound different, readers cannot easily understand how to order or evaluate the responsibilities of the line supervisors. The second example keeps the responsibilities parallel and creates the immediate impression that each has equal weight, and each is critical to the job: *to train, to develop,* and *to support.*

Here are some quick tips designed to keep your writing in line:

Keep elements in a series parallel:

Not: The manager is *young, has little experience,* and *is not very patient.*

But: The new manager is *young, inexperienced,* and *impatient.*

Follow correlative conjunctions (not only/but also, either/or, neither/nor, etc.) with parallel forms:

Not: The new corporate treasurer was known *not only* for keen wit *but* as a financial wizard.

But: The new corporate treasurer was known *not only* as a keen wit *but also* as a financial wizard.

Use the parallel structure with coordinating conjunctions (and, but, yet, etc.):

Not: She likes *to play tennis and running.*

But: She likes *tennis* and *running.*

> Where appropriate, repeat prepositions, verbs, or conjunctions to maintain parallel construction and to clarify meaning.
>
> **Not:** The new account manager was given tips *by* other company executives who liked her *and* the vice president who was her mother.
>
> **But:** The new account manager was given tips *by* other company executives, who liked her, *and by* the vice president, who was her mother.

Parallelism brings more than order to business writing. It brings elegance by creating rhythms and language patterns that have an almost lyrical, musical appeal. Unparallel structures, by contrast, have about as much tonal appeal as an un-tuned banjo.

Solid organization has a strong effect on the reader's opinion of both the writer and the writer's ideas. If you present a logical argument, the reader will more likely pronounce it sensible and support it, even if it is not the best argument. By contrast, if you offer an illogical plan, lacking a coherent structure, the reader will assume it is ill-conceived and unworkable, even if it is a superb plan.

Organization

All writing must begin and end, regardless of its order in between. However, readers have certain expectations of order and completion that the writer must meet to be effective. In general, one basic format serves any approach well: Where possible, have a beginning, a middle, and an end.

The Beginning

Your opening paragraph should draw readers directly into the writing. It should be interesting and inviting, and it should set the rhythms and patterns of the piece that will follow. It should have an attention getter—a come-on that is provocative and promises some payoff—and it should also introduce your thesis or main idea. In sum, the opening paragraph sets up the basis and direction for the rest of the composition by inviting the reader to continue reading and by introducing the central idea, the thesis. Consider this beginning of a brief story about a new editor:

The new editor walked through the newsroom today. After speaking to each of the desk editors and greeting some of the reporters, he went into his glass-partitioned office. He carefully took off his suit jacket and neatly hung it on the coat rack. After methodically arranging his desk, rolling up his shirt sleeves, and sharpening two pencils to fine points, he sat straight as an arrow and began typing like a machine gunner. The reporters, watching this out of the corner of their eyes, began to reread their stories carefully because the editor's reputation had preceded him.

This opening paragraph introduces someone whom neither the reporters nor the readers know—yet. Being new automatically invites curiosity and human inquiry. The reader wants to know more, especially if the editor is as tough as his reputation, which, by the way, is the story's thesis.

The Middle

The middle section should develop the thesis. You may choose your own direction, but in some way you must keep the reader interested and informed. Generally, the middle paragraphs can develop the thesis through opinions, both pro and con, specific details, statistics, or any relative information that will serve to inform, entertain, persuade, or motivate the reader. These paragraphs carry the bulk of the writing, and they must be well done, or they will lose the reader's interest. Make sure these paragraphs are connected with good transition sentences so that the reader is not distracted along the way. See how the following middle paragraphs of the story about the new editor develop the thesis through specific details:

The first column to reach the new editor's desk via e-mail was the social page. Traditionally the sacred cow of the paper, this column had enjoyed editorial immunity for years. But this editor was obviously no timid notary. Fear crept into the hearts of the awaiting reporters as they watched him bear down on his computer screen and begin typing. When he finished, he smiled and hit the send button and grabbed his coffee cup.

Everyone tensely watched the social reporter call up the edits on his computer and read the editorial comments. Soon he began to smile, then openly chuckled. He pushed away from his desk and headed for the cafeteria for his mid-morning cup of coffee.

These two paragraphs clearly support the thesis that this new editor is methodical: we have watched him *bear down on his computer* and *hit the send button*, thus, promptly returning the marked-up copy to the social reporter. So, the new editor certainly seems to be living up to his reputation as a punctual and exacting editor. But why is the reporter smiling at such ruthlessness? The reader must go on—curiosity, you know.

The End

The final paragraph should bring the thesis to some sort of resolution. Remember that the reader has invested time in your writing and is looking for some sort of payoff. Now let's see what happens to our story as we get the payoff:

Gradually, several of the more inquisitive reporters created a reason to drift over to the social reporter's desk. They noticed the page, still up on the computer screen, was covered with red editing notations, but the final comments at the bottom of the page showed a different side of the editor:

Pete, I have always been a fan of yours. You get the scoop even before my wife does, and that is no small task. If I did not edit this closely, however, you can bet we would both get a call from her!

The final paragraph reinforces the thesis that our new editor is indeed tough, but it also reveals a secret: he has got a good sense of humor—and anyone can live with that. Readers enjoy a nice surprise, much like a bonus dividend for their time investment in your writing.

General to Specific

When you want to hit the reader between the eyes, present your conclusion first. By doing so, you develop a show-me attitude among your readers that focuses attention on your central idea or thesis. In effect, you throw down an initial challenge between yourself and your readers. If you can substantiate your up-front conclusion with facts that convince the reader, you will be successful.

As an example, if you were trying to convince a customer to buy a laptop computer, you might start by writing: "This laptop is one of the top five most highly rated laptops on the market." Of course the reader thinks, "Top five, huh? Who says so besides you?" If you can show the reader *a credible authority* who says so, you will have a powerful sales document.

Leading with the conclusion only holds the reader for a short span of time. You have to be able to back it up with quick and convincing facts, or your lead will be overtaken by disbelief and boredom. So, perhaps you could follow up your lead with the following if it were true: "According to *Alpha-Net* and *PC Universe*, this laptop has consistently been the editor's pick for two years in a row." Done deal, assuming these two fictional authorities were real and well respected in the technical world.

Specific to General

Have you ever had to break an appointment? Say "no" to someone? Refute an argument? If you have, chances are you used the specific-to-general approach. This technique generates a persuasive argument and is most effective when you anticipate a skeptical or negative reaction. By gradually building up to the conclusion with fact upon fact, you inadvertently program readers to accept your conclusion. More subtle and less emotional than a direct statement, this technique should be employed for occasions that require letters of refusal and letters of regret. For example:

> Thanks for your invitation to speak at your annual conference. I have high regard for you and your association and have donated to it for many years. Regrettably, because of pressing business demands and travel, I will not be able to speak. Please extend my warmest wishes to all.

This letter of regret will likely get high marks and may even get read to attendees at this conference. It says "no" but with style.

The Chronological Method

The chronological method is popular because humans live in time, remember things as they happened in time, and like to order events in time. This method usually begins with the earliest significant date of an event and works forward to the most recent date. Date sequence, or chronology, provides straight history to the reader and should be used for a background or an overview piece of writing. This technique effectively works when you are creating historical files, documenting personnel folders, or writing progress letters. The chronological method conveys a sense of history and is best used to document the sequence of events.

While generally easy to follow, a chronology should be presented in discreet, logical segments. If you do not limit your time frame, you will lose your reader somewhere in the past. When you are going to talk about the state of modern relationships between men and women, do not start with the creation of man if you expect readers to be awake by the time you get to the 21st century.

In September, we hired two salespeople. In October, we added five salespeople. And this week, we hired three more—bringing the total to ten people on our sales force.

The Reverse Chronological Method

This method starts from the most recent event and moves backward in time. It is best used to jog the reader's memory; you might use it in memoranda when you are trying to remind the boss about something in the past.

We now have ten people on our sales force. You may recall that this past week we hired three salespeople; in October, we hired five; and in September, we hired two.

Caveat on Structure

Remember, structure does not command; it acts as a guide. You can outline a structure and then fill in the details. Or, you can write off the top of your head and then extract your structure. Many writers do not know what they mean until they see what they said. Above all, do not lock yourself into just one formula, or you will be as bored by writing as your writing bores others. But to help bring order and clarity to your writing, use structured approaches like *general to specific*, *specific to general*, *chronological*, and *reverse chronological*, as well as structuring devices like *underlining, capitalization, headings, numbered paragraphs, indentation, columns,* and *bullets.* Look at the following example:

Broken printer:

I am pleased to offer you some solutions to your broken printer. We can refund your money, which means that we could debit your account with us or send the refund to you. We could send you a replacement printer, but would debit it to your account until you return the broken one. Then again, we could send a repairperson to your home, which might take several weeks. You may want to consider these options. Please let us know and thanks for your patience.

Fixing the printer:

To fix your printer problem, we can do any of the following:

- Refund your money. We could debit your account with us or send the refund to you.

- Ship you a new printer. We would not debit it to your account until you return the broken one.

- Send a repairperson to your home. This might take several weeks.

Please let us know your choice, and thanks for your patience.

A final word on order: Don't let order or structure become your master. The best advice might be the adage, "All things in moderation."

CHAPTER 7

WRITE WITH STYLE

Most inexperienced writers think that great authors receive eloquent language from some divine force in the universe. Great authors, they think, sit entranced at their desks as well-polished sentences flow from their fingers. Fact: writing is a *skill*, learned and refined through practice and hard work. If, then, an experienced writer works easily because of constant practice, inexperienced writers should react to their own difficulties by practicing—not by mumbling about God-given gifts.

The Writing Process

Writing is not a one-time or sudden event, but rather a process—a series of steps necessary to achieve a finished piece of effective writing. Collectively, these steps make up the process of writing:

Selecting

First, you must select an idea that will interest you. This idea should readily link with you previous experiences, ideas, and interests, and it should strike a nerve deep within you. In other words, you must select an idea that has enough meaning to you that it will stimulate your interest for a sustained period of time. Of course, in business, necessity or your boss might well dictate which topic you select.

Exploring

Once you have selected or been assigned the topic, give it to your unconscious mind for a day or two—if you have the luxury of time. Stranger than fiction, this unconscious technique is an effective one. By choosing a topic firmly, then ignoring it, it will rattle around freely in your mind and link up with connected, but forgotten, data in your brain's deep storage. Soon these unconscious fragments will begin to surface in your conscious mind. At the most unexpected moments, you will be struck by relationships between new ideas and old experiences. As they occur to you, record these relationships as notes, in outline form, or even as long, windy paragraphs. Sometimes these ideas will work and sometimes not, but they will always give you a good place to start. This step explores that marvelous computer called your brain. In writing, a scan search of your unconscious mind takes advantage of your total life experiences.

Scrawling

Now for some fun: Sit down at your computer or with a pad of paper—whatever is your preferred medium—and type or write anything that comes into your head regarding your topic. Don't stop or hesitate for neat typing or correct penmanship. Don't stop to correct spelling, to punctuate, or to complete sentences; just get your ideas on paper. I call this step "scrawling" because it is somewhere *beneath* writing, and it describes the kind of free-flowing, free writing activity that I hope you will have the courage to try. Once you have gotten the hang of it, you'll enjoy it, and you will discover you have more and better ideas than you ever would have guessed. Remember: Your brain is an amazing tool. However, if you are used to editing your writing word by word or line by line, at first you may find it difficult to scrawl. Letting a misspelled word or an incomplete sentence stand until the whole idea gets laid out is most difficult, at first, for some people. However, you will benefit from this technique if only because it yields concrete forms from vague ideas. Think about it: If you had to fix a faulty widget, would you rather have the widget in your hands to feel and inspect or hear a vague, general description of a widget?

Surveying

Now, survey what you have written and begin to look for informational holes—places in your scrawling that lack depth or substance. Edit the scrawling first on a grand scale. Label similar groups of ideas with the same letter or number to help you establish how much information you have in a certain area. This technique will show you graphically and quickly where you need work. For example, if you only have a few letters or numbers in an important area, you will need more research there.

Hunting

To fill the holes you do find in your writing, hunt out books, magazines, papers, and documents. Go to colleagues, subject matter experts, and others to find the information you need. You may realize that your idea is too large or too small for your particular purpose. You may find that you will have to limit your topic, widen it, or in some way reshape it. If you take the time to hunt properly, you will define your limits now and prevent a lot of wasted time and writing later.

Writing

Once you have gathered all your documentation, start writing. Again, write quickly to get the ideas on paper. Because you are armed with facts this time, your writing will be more substantive. The result will be your first draft and require lots of editing, so leave wide margins all around and double, even triple, space this draft. When you finish the draft, let it sit for a day or two if you have the time. Time gives you objectivity and enables you to see your draft with the cooler eye of an editor.

Editing

Edit your initial draft to get a more concise, accurate, and reliable expression of your ideas. Scribble notes up and down the page; cross out phrases and entire sentences; draw arrows over, around, and through words. Then put your writing aside for awhile—again to get a little distance from it. You need the distance to clear your mind and to prepare you to make more changes.

Rewriting

Rewriting is as arduous as it is productive. As a rule, the quality of writing improves in proportion to the number of times a piece gets rewritten. Nevertheless, both the editing and rewriting processes are painful because they force you to give up writing that you like but that does not fit. Saying goodbye to well-crafted phrases or witty ideas isn't easy. Also, remember that the longer you work with a piece of writing, the harder it will be for you to give any of it up.

Testing

Now, test your writing with people whose opinion you respect—the more people the better. Be careful not to over-prepare or influence the members of your test group before they read the piece, and tell them to be honest, direct, and critical. When you get back their edits, you decide on what stays, goes, or needs to be rewritten. Often neglected by new writers, this step is seldom forgotten by experienced writers. Repeat this step after different drafts and as often as possible before the deadline.

Ending

At some point, you have to finish the writing to meet a deadline. If there were no deadlines, many writers would rewrite indefinitely. Rewriting is only limited by your patience. However, as a rule of thumb, when your writing begins to come back with few or no editing remarks from other readers, it is probably ready for the final draft, proofreading, and publication. Don't be surprised if you have mixed feelings when you submit it.

Tips for Successful Writing

Now that you are aware of the writing process *in general,* consider some practical tips for writing that gets the job done at work:

Know What You Want

Take a moment before you begin and reduce your writing purpose to a simple sentence: "I want this customer to know that our company is sorry for our mistake," or "I want our manufacturer to know that we are upset he missed our deadline because it cost us money." By doing this, you will clarify your purpose and stick to the mission at hand. This purpose sentence gives direction to your writing.

Respect the Reader's Time

Businesspeople are pushed for time. They have many and con-stant demands placed on them. Don't contribute to their stress by drafting a communication that requires a $100 investment for a 10-cent problem. Keep your writing clear, concise, and to the point. If not, chances are it might just end up at the bottom of the pile or, worse yet, unread in the trash can.

Keep the Reader's Needs in Mind

If you want your writing to be understood, always keep the reader's needs, demands, concerns, and point of view in mind as you write. Avoid unfamiliar language or terms, and address questions you know the reader will logically have about your subject. In short, follow the golden rule of writing: Write for others as you would have them write for you.

Speak Directly to the Reader

To avoid stilted and formal language that can put off the reader, pretend that you are speaking directly to the reader. A tape recorder may be useful to help you get started on an early draft. Be careful, however. Don't forget to edit your conversation strictly because transcribed speech, without the benefit of accompanying gestures, can be too informal, too chatty, and too sloppy.

Lead with the Main Point

Lead with your strong point. Don't make the reader go on a treasure hunt to find your gem. Don't bury the main point in the middle of a paragraph, or the reader might miss it. To discover what your main point or main sentence is, read a paragraph. Now decide which sentence would be the very last one you would toss out. Use that sentence as your main point of the paragraph and lead with it.

Make Writing Easy to Read

Make your writing pleasant to read by using some simple techniques:

- Underline for extra <u>stress</u>
- Use lists and bullets
- Do not fill the page with typing; use white space to rest the reader's eyes. For example, skip lines between paragraphs
- Use subheadings to break space and ensure clarity

Think of things you have read that appeal to you. Save those examples, study them, and use the devices you like in them.

Don't Overwrite

Consider the impact of the Declaration of Independence, the Preamble of the Constitution, the Lord's Prayer, and the Gettysburg Address. All of these written pieces are short but powerful. Don't be afraid to cut.

Keep It Simple

Never use two words when one will work nor use a paragraph when a sentence will work. Strive for brevity and simplicity, especially in business writing.

Don't Write Alone

Always get someone else to criticize your writing objectively before you submit it. The more you work with your writing, the less objective you become; so ask someone else to read specifically for clarity and accuracy. The only thing better than one person editing is two people editing. You don't live in a vacuum, so why write in one?

Avoid Injecting Your Opinion

Stick to the facts in business writing. Rarely does the reader want your gratuitous opinion on a particular matter, especially if not rooted deeply in facts. Deal with facts and avoid the unmistakable mark of egotism in your writing.

Avoid Qualifiers and Vague Modifiers

Don't use modifiers that almost say what you want when you can find ones that are exact.

Close	Exact
The shirt was very expensive.	The shirt cost $300.
The train was very late.	The train was four hours late.

Don't Use Clichés

Worn-out expressions can wear a reader's patience thin. Avoid hackneyed phrases and use direct, fresh language:

Trite	Natural
First and foremost	First
All around the mulberry bush	Everywhere

Avoid Exaggeration

Avoid the superlative when you write. "The greatest," "the worst," "the prettiest" all leave you open to exception. To avoid leaving arguments in the reader's mind, avoid over-stating your position.

Sum It Up

The last paragraph should summarize what you have been discussing. Also, a summary paragraph should tell the reader what you want done or what you will do. It's a call to action.

Speaking of summing it up, don't forget these tips on writing:

- Know what you want
- Respect the reader's time
- Keep the reader's needs in mind
- Speak directly to the reader
- Lead with your main point
- Make writing easy to read
- Don't overwrite
- Keep writing simple
- Don't write alone
- Avoid injecting your opinion
- Avoid qualifiers or vague modifiers
- Don't use clichés or trite expressions
- Avoid exaggeration
- Sum it up

CHAPTER 8

LEARN HOW TO BECOME A GOOD EDITOR

808. Clle AZ779w
808. Dlee Guttn

Woriens wach wards
un busmuns 2, le geel
2nstrc.

Agnes Lynn

Mighter than the Sword
and -

			Call number	Author	Title
✓	mncir	y	914.15 K2918f 2003	Kelleher, Suzanne Rowan	Frommer's Ireland 2003
✓	mncir	y	914.160824 F613s	Fletcher, Martin, 1956-	Silver linings : travels around N
✓	mncir	y	914.21 C7372		The complete book of London
✓	mncir		914.3613 V662 2002		Vienna
✓	mncir		914.39 B9271		Budapest
✓✓	mncir		914.4361 P2321 2004		Paris
✓	mncir	y	914.4361 P8446r 2002	Porter, Darwin	Paris 2003 : selection d'hotels
✓	mncir		914.4361 P2321m 2003		
✓	mncir	y	914.604 R856 1997	Ellingham, Mark	Rome
✓	mncir	y	914.8 P8446f 2001	Porter, Darwin	Spain : the rough guide
✓	mncir		914.94 S599s 2003	Simonis, Damien	Frommer's Scandinavia
✓	mncir	y	914.95 R49e	Rice, Tamara Talbot	Switzerland
✓	mncir		914.955 E2665r 2003	Edwards, Nick	Everyday life in Byzantium
✓	mncir		914.9618 T2433i	Taylor, Jane	The rough guide to Corfu
✓	mncir	y	915.694 F6532 2001		Imperial Istanbul : a traveller's
✓	mncir		917.299 H312m 2004	Harriott, Catherine	Fodor's Israel
✓	mncir	y	917.3 H23687a	Handlin, Oscar, 1915-	Maverick guide to Bermuda
✓	mncir	y	917.3 S81t 1997	Steinbeck, John, 1902-1968	American people in the twentie
✓	mncir	y	917.4 M6875m 2002		Travels with Charley : in searc
✓	mncir	y	917.504 B5162a 2002	Berman, Eleanor, 1934-	Mobil travel guide : Mid-Atlanti
	mncir	y			Away for the weekend : mid-A

Eventually, all writers have to learn to edit—both their own writing and that of others. It comes with the territory. In fact, most writers won't improve their writing significantly until they start editing—using critical thinking skills. However, critiquing the writing of other writers not only requires skill but also diplomacy. Editing can be a traumatic experience for both writer and editor. Though writers may tell you they don't mind your changes, you can be assured that they well may mind any changes. Because writing expresses an individual personality, writers take personal notice at suggested changes. Thus, editors may see their well-intended remarks rebuffed, even maligned, and wonder why they bothered to help.

General Guidance

Before listing some specific editing techniques, let's look at several general guidelines about the process.

Be Gentle

Use unemotional words unless you are praising the work. Don't use words like *awful* or *terrible*. Treat writing as if it were the writer's child. Use phrases like *please clarify* or *expand this idea*. If you take out the emotion, tension will stay low and productivity high.

Use P-Q-P

First, offer praise (P); tell the writer what you like about the writing. "I like the way you included such rich detail throughout the piece." Second, ask questions (Q); ask the writer to clarify certain questions that the writing raised, or ask questions to stimulate the writer's thinking, such as: "What do you mean when you said 'things'? Can you be more specific? Or, did you consider describing the incident in more detail to make it come alive?" Third, offer polish (P)—that's polish like what you do to furniture, not Polish, as in from Poland! Help the writer by offering specific help to polish the writing even more, such as: "You might want to consider adding an even stronger attention-getting opener to your piece." Using P-Q-P ensures that as an editor, your comments will get the respect and attention they deserve. You'll also provide criticism that will help and not destroy the writer's confidence.

Work Collaboratively

Working together, the editor and the writer form a powerful and effective team. Keep the editing process collaborative and enjoy the fruits of mutual success.

Use a Stylebook

A number of excellent writing stylebooks are available on the market. Use the stylebook of the *Washington Post*, the *New York Times*, or some standard, and use it throughout your organization. Be sure everyone has a copy and insist everyone comply with it. Conformance to a standard stylebook will give your business a distinctive and effective writing voice.

Editing Tips

When you edit someone else's writing, remember these simple but effective techniques:

Edit for Clarity

All writers have their own particular style. Edit style only if clarity is at stake.

Edit for Conformity

As a reviewing official, you must understand and enforce company practices, standards, and policies. To ensure conformity, closely review the writing of newer, less experienced employees.

Edit for Tone

Extract emotionally charged words from the text. Keep the tone of business communication forthright, reasoned, and professional. Emotional writing invites overreaction and causes problems.

Edit for Organization

Readers, being human, understand and expect logic. The structure of a piece of writing must, therefore, organize and sustain the ideas in it. Edit structure to ensure that the writing in it will be understood by the greatest possible number of people.

Edit for Fact

Challenge questionable matters of fact because you can be sure readers will. Both writer and editor must guarantee the accuracy of facts.

Standard Proofreading Symbols

The following proofreading symbols were devised as short-form directions to the printers. They are standards and will make your job editing infinitely easier:

Symbols	Results
Delete this error.	Delete this error.
Place a comma here please.	Place a comma here, please.
Use a colon for the following	Use a colon for the following:
Always capitalize the capitol.	Always capitalize the Capitol.
Use figures for numbers like thirty.	Use figures for numbers like 30.
Insert an apostrophe when its needed.	Insert an apostrophe when it's needed.
Indent new paragraphs.	Indent new paragraphs.

(continued)

Symbols	Results
Insert mi*s*ing letters.	Insert missing letters.
Link words like news⌐paper.	Link words like newspaper.
Join letter⌐s to make sense.	Join letters to make sense.
Transpose <u>words</u> these.	Transpose these words.
Link this copy⌐ and what follows.	Link this copy and what follows.
Transpose these lette*rs*.	Transpose these letters.
Necessary Insert words.	Insert necessary words.
Use boldface capitals. **BFC**	Use **BOLDFACE CAPITALS.**
Use lowercase letter Here. **LC**	Use lowercase letter here.
Insert a period⊙	Insert a period.

(continued)

95

Symbols	Results
Spell the name Joans⟨cɐ⟩ as written.	Spell the name *Joans* as written.
Spell out numbers ① through ⑩.	Spell out number one through ten.

CHAPTER 9

MORE ABOUT
EDITING

When most people are asked to edit a piece of writing, they have difficulty knowing exactly where to start. So, consider approaching editing from two different levels: macroscopic (BIG STUFF) and microscopic (small stuff).

Please note: We've previously discussed a number of the following topics, but now regard them as an editor, not simply a writer, because a good writer should also become a good editor.

Macroscopic Editing

Macroscopic editing looks at writing from an overview perspective. It hovers above the text and considers its overall strengths and weaknesses. What follows are some macroscopic tips for editors.

Check the Basic Structure

When you edit a text, ask yourself if the piece has a beginning, a middle, and an end—the basics of rhetorical structure. Does the writer tell the reader what she'll tell him (introduction), then tell him the details (the body), and then tell him what she told him (the conclusion)? Also, did the writer employ any rules of proportion to the text? Typically, the introduction consumes 15 percent of the text, the body about 75 percent, and the conclusion about 15 percent of the text. Thus, in a 10-page document, 1.5 pages will be dedicated to introducing the topic; 7.5 pages (broken into several main areas of relative proportional length) will compose the body of the text; and 1 page will do it for the conclusion.

Position the Bottom Line

When a piece of writing offers good news, make sure the writer puts her best foot forward: Put the bottom line first. "Our team reached all our goals this year. Specifically, we. . . ." Should the writer have bad news to report, make sure she delays the bottom line so that the reader has time to get used to the idea first. "Last year our team was cut by 33 percent, and our operating budget was cut by nearly 50 percent. The result of such cost savings had an understandable but unfortunate impact on our revenues as well."

Cut Long Sentences

Scan the draft quickly and look for more than three lines of copy without a terminal piece of punctuation: a period, comma, or question mark. Usually when you do find such sentences, you'll often find confusion that can stand editing and simplifying. Remember the Rule of Thirty we discussed previously.

Insert Paragraphs

While you're scanning the draft for lengthy sentences, also scour it for pages without paragraph breaks, because long blocks of unorganized text confuse readers. As a rule of thumb, a good paragraph has a lead or topic sentence that links itself to the preceding paragraph, offers an overview of the details to follow, then offers several sentences with details supporting the topic sentence. Finally, the last sentence of a paragraph leads to the next paragraph.

Use Transitions

To keep readers on track, link sentence to sentence and paragraph to paragraph. Here are some ways:

A coherent text can be linked by *repeated words.*

The black *Corvette* was parked at the curb.
Approaching the *Corvette,* Jack walked slowly.

You can link text by using a simple *pronoun.*

A man approached the Corvette. As *he* got close, he stopped.

Finally, you can use *transition words* to signal your direction to the reader. It's like turning on your blinker on the highway to let the car behind know where you're headed.

Approaching the Corvette, Jack walked slowly. *When* he got close, he stopped. *Then,* he stared into the driver's window, only to find out there was a mannequin in the driver's seat.

Check Voice and Diction

Who narrates the action in your writing (voice) and the words the narrator chooses (diction) make a big difference in the presentation.

One Voice:	When Jack came in, the place went quiet enough to hear a pin drop on the floor.
Another voice:	When Jack came in, I could hear a pin drop on the floor.

When you make word choices (diction), that language can be loaded emotionally. Consider these two sentences for differences in diction:

Sentence 1:	Jack exploded into the room like a hand grenade.
Sentence 2:	Jack came into the room following some yelling and banging on the door.

Try Horizontal Editing

Sequentially lay out the paper on a desk or some accommodating surface (horizontal editing) and look for proportionality throughout the text. Roughly calculate how many paragraphs or pages were used in each section. Often writers spend a lot of time and space writing about what they know, but lightly skip over un-researched sections. But it's easier to see proportion in a text when you physically lay it out horizontally on a surface, which allows you to view text stretched out before you.

Edit Yourself

When you must self edit, try these tips: Give the writing time—set the draft away for an hour or a day, if possible, to give you a fresh set of eyes. Read it out loud. Wherever you stumble, you'll usually find the writing faltering.

Reformat Your Document

Here's a great technique. Change the page layout from one column to two columns. The new formatting actually tricks the brain and makes it think that the writing is new, allowing you to review it with fresh eyes.

Microscopic Editing

Microscopic editing looks at writing from an up-close perspective; this approach considers each word's overall strengths and weaknesses. Many of these tips were mentioned earlier in this book when discussing good writing, and they are critical to good editing as well. Here are some microscopic tips when you have to edit:

Use Action Verbs

As mentioned earlier, action verbs create strong pictures. So edit in action verbs.

Not:	The CEO seemed upset.
But:	The CEO scowled, banged his oak desk, and yelled, "No!" at the sales manager.

Avoid the Verb *is*

Scour the text for overuse of linking verbs such as is (was, were), seem, appear, look, taste, smell, and others. These verbs lead to using adjectives that often lessen the impact of the image. Instead, inject a more active verb. Start by locating all the *is*'s, *was*'s, and *were*'s. See if you can replace them with active verbs.

Not:	The day was gloomy.
But:	Clouds covered the sky, and rain pelted the street.

Keep Related Words Together

This was mentioned before, but bears repeating especially for editors. When you find modifiers in the text, move them close to the words they modify.

Not: Crying and stomping his feet, the manager observed the young boy on his security monitor.

But: Using his security monitor, the manager observed the young boy crying and stomping his feet.

Put Important Words First or Last

People remember what they hear first (rule of primacy) or what they hear last (rule of recency). So if you heard a string of nine numbers. . . . 7, 3, 2, 5, 9, 1, 8, 4, 6. . . . and then were asked to recite the first or last numbers, you'd likely be able to. However, if asked to recite the middle number, you'd be hard pressed. Edit accordingly.

Not: The girl gave the president a kiss on the cheek.

But: The girl kissed the cheek of *the president*.
Or: *The president* was kissed on the cheek by the girl.

Note that while you may have to use the passive voice in cases like this, it's worth the tradeoff to emphasize *the president*.

Avoid Shifts in Number Tense or Point of Voice

This was discussed earlier. Most of us view consistency as desirable and inconsistency as untrustworthy. Take care when editing.

Not:	The workers eats his lunch.
But:	The workers eat their lunches.
Not:	I walk into the executive offices, and I got my bonus handed to me without a word.
But:	I walk into the executive offices, and the boss gave me my bonus without saying a word.

Avoid *it is* and *there are*

Such openings are common, but can detract from strong writing.

Not:	There are a lot of reasons why the CEO fired the managing director.
But:	The CEO had a lot of reasons to fire the managing director.

Use Proper Punctuation

Punctuation—especially commas—cause people problems. Here's a list of comma rules to help you:

- After an introductory element like this, use a comma.

- Commas should set off additional information, which might help the reader, but may not be necessary.

- Use commas in a series between words, phrases, or clauses.

- Commas belong between separate equal, equivalent modifiers.

- Commas are used, as you might suspect, to set off parenthetical information.

- Commas set off contrast, not similarity.

- Jack, use a comma in direct address.

- "Use a comma with quotations," he said.

- Use commas with numbers (2,700) and dates (July 1, 2005).

- Finally, sparingly use a comma to avoid confusion, rather than just using it willy nilly.

Check Out Some Fun Editing Tips

- Avoid not using no double negatives.

- Check speling.

- For what it's worth, avoid clichés at all costs.

- Be careful not to inadvertently or otherwise split infinitives.

- Avoid jargon ASAP. Roger that.

- Run-on sentences require punctuation they're awkward otherwise and confuse readers they deserve better.

- Abbrs. should be kept to min . . . OK?

- "Use a comma with quotations," he said.

- Use a dictionery.

- A writer should keep his language gender neutral.

- I should not shift your point of view.

- Parallel construction, helps clarify, to organize and ordering.

- Sentence fragments.

CHAPTER 10

WRITE TO YOUR AUDIENCE

Writing communicates ideas between people. But it is not until someone reads the writing that the process is complete. Despite this, readers are often forgotten by writers. This chapter will profile the audience of written business communication and define the kind of relationship that should exist between the writer and the readers.

Put People in Your Writing

People do things. People take action, respond, and comply with regulations. And companies can be thought of as gatherings of people acting together to produce a product or service. Unfortunately, when companies write to other companies, they use sterile language that dehumanizes the communication— makes it cold, lifeless, and boring. It's incredible when you think about it. We try to inform, explain, or motivate, yet we linguistically approach the situation as if it were a rare disease requiring a surgically sterile isolation. We put on masks, sterile gloves, and gowns to hide ourselves, and even drape the patient (our writing) so that both patient and surgeon have completely lost their identities. Then, we attempt to enter into one of the most intimate acts: communication.

But when we speak to people, we unconsciously use a very personal touch. We address the person, not the company. We get to know people a little before we make demands of them. We assimilate knowledge about them that helps us shape our communication approach to meet their needs. We look at how they dress, and we listen to what they say and how they say it (dialect, intonation, diction). In fact, we absorb a remarkable amount of information about an individual before we ever speak a word. The entire process only takes seconds, but it happens nonetheless and crucially effects good communication. In general, the more time you have to assess your audience, the better chance you have to make your point.

However, in writing, we often ignore the reader and focus on our theme—*our* main point. But what *we* want is not necessarily what the reader either wants or needs. Here, writing fails. Consider how easily the problem can be corrected:

Write to a Person

Avoid addressing a letter to "Dear Sir," "Gentlemen," or "Dear Madam." These people are all dead—in fact, they never lived. They are ciphers, nonentities, and they dehumanize the person who receives the letter. Take the time to find out a name and send your letter to a real person: "Dear Mr. Johnson," "Dear Harry," "Dear Somebody."

Be a Somebody

Let the reader know you are a real person with a name, a sense of humor, and ideas. Relate to your reader in human terms just as you would at a social gathering. Your language should reflect your personality. Tell stories and relate interesting situations, especially humorous ones, to illustrate a point. Do not shy away from using words like *I, me,* or *we.* But avoid referring to yourself as "the company"; the company is an "it."

Assess Audience Needs

Most people will accept or do something that has a payout for them. Motivation begins at home. If you have done your homework before you write, you should have some idea about what motivates your readers and what does not. Armed with such intelligence, you can write to their needs and be sure of a better reception.

Know What You Want

Decide in a nutshell what you want your communication to accomplish and what action you want readers to take. Then, ask them in plain English. Call your readers to action. Ask them to fill out a survey and return it, to visit your company's Web site, to call you—to DO something you can measure.

Not:	Have a great day.
But:	Call me at 1-800-876-5432 for more information and a 30 percent discount on your first order.

Use Appropriate Language

Just as you would never send a letter written in English to a reader who does not speak the language, you should also never send a message to an audience written in high-brow language they may not understand. Avoid flowery and pretentious language. Use language that your readers use. You can find out the appropriate style by reading and studying some of their correspondence. Become a bit of a detective. If you have no writing samples, make notes the next time you speak with them. You will hear idioms, rhythms, and regional expressive style after which you can pattern your own writing. If your readers identify with you, even stylistically, they will give your proposals greater consideration.

Not:	Your gracious consideration of our proffer will be punctually acknowledged.
But:	Thanks for considering our offer. We'll respond to your decision promptly.

To sum up, these initial steps will help put people in your writing:

- Write to a person, not a company.
- Identify yourself as a person, not a company.
- Assess and meet the readers' needs.
- Know what you want.
- Use appropriate language.

Audience Types

How many audiences are there? As many as there are people and groups of people. In business, however, correspondents fall into two general types: internal and external. Internal audiences are those people inside your organization to whom you must communicate: superiors, peers, and subordinates. External audiences are those people outside your organization to whom you write: consumers, vendors, competitors, and the news media.

Writing Inside the Organization

Let us begin inside before we venture outside.

Writing to Superiors

Communicating with your supervisor is essential to your survival in the company. Many of us must communicate daily with the boss on a number of pressing subjects. How well we communicate our ideas to her or him—whether in written or spoken language—determines our success in the organization.

When writing to supervisors, keep in mind the general tips on writing: What's your relationship? How formal or informal is the relationship? What do they like? Do they have a sense of humor, and if so, what type? The better you know your superiors, the better you can match your language.

Tips for Writing to Your Supervisor:

- Ask a third party who knows your boss to read your memo and to give you a reaction.

- Enlist your boss's assistant or staffer who may give you valuable insight into her moods, needs, and wants.

- Read your boss's most recent memos to get a sense of diction, style, and tone from them.

- Look at your boss's bookshelves and at pictures or sayings hanging on the wall of his office. These can give you good personality clues that you need, especially if the boss is new and not well known.

Writing to Peers

Communicating laterally in an organization often presents a particular challenge because we have been conditioned to compete, not communicate, with our peers. Like sports, business presents a competitive arena, and after all those years on the athletic field, we are trained to beat the other person, to seek the advantage, and always to keep score. However, if you don't moderate that competitive, win-lose approach with your peers, you'll find yourself on the losing side.

Tips for Writing to Peers:

- Get to know your peers. Establish a relationship with them *before* you enlist their aid.

- Remember: Friends help friends.

- Keep your language direct and forthright.

- Establish what you want and how getting it will benefit your peers. Your peers will react negatively to a sales job. Hit them straight with what you want to avoid arousing their competitive instincts.

- Work collaboratively with peers. If you can persuade potential competitors to join you in a project, it will be doubly successful; you will have their support instead of their resistance.

- Finally, offer alternatives in case your first idea is rejected.

Not: Please endorse my new plan to redistrict the sales territories.

But: This new sales redistricting plan gets you an extra salesperson and keeps my district manager from having to travel so much.

Writing to Subordinates

What's the bottom line in business? Simple: To make a profit. Nevertheless, the way managers motivate subordinates to do the job determines which managers are successful. Managers write to subordinates for two primary reasons: (1) to set out a task and (2) to document instructions. Because most forms of performance rating are based on written documentation, managers must put tasks and instructions on paper. However, managers also have to build relationships to help make those tasks get done easier and more effectively.

Tips for Writing to Subordinates:

- Write collectively. Include the subordinates on your team. Use words like *we, us,* and *our.*

- Instill a sense of corporate ownership. Convince subordinates that their role is integral to the overall performance of the company.

- Be polite, not patronizing. Simple words of courtesy help get the work done. Thank subordinates for their work, ask for their help, and seek out their ideas.

- Write out praise. If subordinates have done a good job for you, let them know it in writing. Be specific with your praise so that you do not sound like an insincere backslapper.

- Tell them when you are not pleased. The sword cuts both ways, but when you criticize, be specific and address only specific acts, not general personality traits. Specifics help employees focus on and correct errors and avoid taking criticism personally.

- Use humor; save the somber stuff for those times that demand it.

- Be creative in everyday writing. Short memos and notes to employees can be creative. Don't miss the opportunity to exercise your mind and your pen. Try to find different ways of saying things. Try a new style or use a new word just to practice it. But, keep it simple and direct.

Not:	Get the report in by noon today.
But:	Jack, we're running up against the proposal deadline. Please get it to me by noon today.
Not:	Jack, I'm very disappointed about the deadline.
But:	By not meeting the proposal deadline, our company did not get the contract. I'm not pleased and want to discuss it with you.
Not:	Thanks.
But:	I appreciate your work on the Smith account. The research was especially superb. Thanks.

Writing to subordinates can be an indispensable managerial tool. Too often we avoid writing and miss that opportunity.

Writing to the Public

As you climb the corporate ladder, you'll write increasingly to people outside the company. In fact, the higher you climb, the more you will affect the world around you, and the more you will need to correspond with that world. Of the numerous external audiences with whom you might have to correspond, here are just a few.

Customers

Customers are surely one of the most important groups with whom you communicate. Without them, business does not happen. When responding to consumers, be direct, forthright, and accurate; ensure that answers to specific problems are correct and courteous. Be sure not to use patronizing language, or you might lose your most valuable asset, a cash-paying customer. Avoid impersonal answers and remember that the consumer is a person just like you.

Not:	Company policy demands payment on purchase.
But:	By policy, our company requires payment on purchase.

Vendors

Suppliers of the daily goods necessary to your company are also vital and should be treated personally. When writing to them, remember that even though, like employees, they are paid for their services, they also need to be told when they do well and when they do not perform. Writing provides the occasion and means to express pleasure or displeasure. Of course, your diction may vary, depending upon your relationship with the vendor. Certainly, your diction may vary from dealing with a new vendor to one with whom you have been associated for years.

Not:	Goods received.
But:	Susan, we got the shipment on time and under budget. Thanks, again.

News Media

The Fourth Estate, otherwise known as the news media, can make or break your company. Handling the media has become a well-defined skill necessary to corporate survival in today's news-hungry society. How your company responds—especially during a crisis like a strike, a product recall, or a disaster—may be the public's only view of your operation besides buying your products. The image you project will determine if your public likes or dislikes you, which can mean a swing in profits. Always prepare before you meet the press. Write out in detail well-considered responses to anticipated questions. Have someone inside the company play the devil's advocate and ask you tough questions. Rehearse your responses. Spontaneity is a poor substitute for solid preparation. Make sure that your responses are honestly what you believe, and then rehearse your delivery. And, avoid "no comment" like the plague. If you can't comment, tell the media why. Remember that your delivery in this case is as important as your written preparation. When reporters take notes or when the cameras and tape recorders whir, make sure you're prepared. Remember to keep a positive tone to all communications and consider the media's need for information.

Not: The union is being completely unreasonable during negotiations, so we just can't make any progress.

But: Despite slow progress, we are negotiating in good faith with our union and are confident that we can come to some agreement.

Not: No comment.

But: Unfortunately, I can't comment on the specifics of that situation due to advice from counsel. However, our company is working to resolve that issue within the week. I'll get back to you as soon as we do resolve it.

External audiences are varied and many. All of them have some bearing on your company's health and welfare and, therefore, must be effectively communicated with. When writing to any such audience, whether as big as the government or as small as a one-person business, writers compete for a reader's valuable time. To be a winner and an effective communicator, a writer must strike a bond with the reader. To create this bond, it is the writer's job to do some background investigation on the audience well before putting pen to paper.

CHAPTER 11

WRITING E-MAIL

Without a doubt, the most potent application of the Internet has to be e-mail. It's not only the way that friends keep up with each other, but also has become the backbone of business. Both internal and external business transactions take place with amazing speed and efficiency by using e-mail. However, you can easily and inadvertently step on someone's cyber toes and as a result damage your credibility quickly if you're not careful. So look at these simple observations and tips about e-mail before you trip over an e-mail faux pas.

E-mail in General

Here are some general observations about e-mail.

Fast

With the click of the send button, e-mail is gone. That's good and bad. When replying to an e-mail, take time before you click the send button and carefully read to whom you're actually replying. Sometimes you may not want your message to go to everyone still left on the reply line. In fact, be careful that you send the e-mail only to those you want to receive it. Check the "To:" line very closely every time before hitting the send button. You would not want to send a very personal e-mail to a business acquaintance simply because you miss-clicked in your address book.

Note the difference in e-mail between "reply" and "reply all." "Reply" sends a response back to the originator of your incoming message, regardless of how many people may have been copied. "Reply all" sends your response to everyone—including the long list copied on the original. Often you may *not* want to do that. Bottom line: Take care before you hit the send button. Always.

Sloppy

Some people tend to treat e-mail like a fast note and not a well-thought-out, refined document. As a result, sometimes they send sloppy writing to people who matter most to them or their business. Take heed to send informally written e-mails to friends and more carefully written (and edited) ones to business associates and customers.

A Broadcast Medium

Unlike a letter, an e-mail can be forwarded by the recipient to thousands of people, posted on a user group, or stored for a later date—in some cases forever. This makes e-mail both a more potent and more dangerous form of communication. Be careful what you write in your e-mail because you may have little control in who ends up reading it.

Feels Private

In most systems, the network administrator has full ability to read every e-mail you send, so be careful. Some companies monitor employee e-mail for security and control issues. While you may find that draconian, it's a reality. Also, people can forward what you said, and software can go fluky and send your stuff to crazy destinations you never planned. Remember that there is no such thing as privacy on e-mail.

Copyrighted

Since 1990, anything people write is considered copyrighted. Thus, especially in formal situations or where you don't know people well, you should get their permission before forwarding. As a courtesy, treat e-mail like a book draft: Quote it and get permission particularly since you never know yourself where it will ultimately end up.

E-mail Do's and Don'ts

Here are a list of do's and don'ts to save you unnecessary heartburn when writing e-mails.

Keep it Brief

E-mail's top advantage remains its speed and conciseness. Don't squander away this advantage with long, rambling messages. Doing so insults the reader and gets your e-mail filtered out to a file in obscurity or printed out and put in a hold bin for later reading that may never happen.

Respond to Messages

Try to be responsive to e-mail requests, but take time when they're important. Try to respond daily to e-mails, but it's perfectly fine to take a day or two to think about what you're going to say, particularly if the results could help or hurt you. If need be, consider an interim response like, "I'll get back to you in a few days when I've had time to think through the options."

Don't Write with Emotion

Emotions can make you say things quickly that you may not want recorded. For example, if you have finally gotten a marginal employee to actually do something, and you're elated, you might be tempted to say overly complimentary things. Be careful: You'll see that effusive praise at a performance meeting where you might be trying to dismiss that person. On the other hand, when you're angry and are tempted to fire out an emotionally charged response, don't. You'll regret it— guaranteed. Write all you want to vent your frustration, but save it as a draft document and wait at least a day or two. You'll likely never send such a flaming document and save yourself endless grief.

Stick to the Point

Keep e-mail focused on one topic and if possible a single screen. Briefer is better in this flash-like medium. Just as short memos tend to get read first, the same is true with e-mail. If the reader can see the start and finish on one screen, your e-mail's chances for getting read increase exponentially.

Avoid Abbreviations

To communicate faster, e-mailers have developed their own abbreviations like BTW (by the way), FWIW (for what it's worth), TTFN (ta ta for now), TNSTAAFL (there's no such thing as a free lunch), and so on. These can be endless and annoying, especially if the reader doesn't know the code. Use them sparingly and only with folks you know will understand you.

Frown on Smilies ☹☺

If you use these little characters to express emotion ☺ you might make people scream :-O. They're fun and express emotion, but they can be distracting and childish, and detract from your message (especially when people don't know the convention). If you want to use them, send them only to friends ☺, not in business writing ☹.

Use Quotes When Responding

Don't just send a message back that says "OK with me" without some context for the recipient who may have sent out 100 e-mails that day. Offer some quote back for context, not necessarily the entire message. And answer at the top of your message, not the bottom, to avoid requiring readers to sort all the way through the piece. Here's an example:

> Joe,
>
> I agree. Buy the 1,000 shares of ABC.
>
> >If you agree with the proposal that we buy 1,000 shares of stock in ABC corporation, let me know ASAP. >

Avoid Flaming

Flaming equates to an online shouting match. Some simple advice: Don't write angry messages. And to avoid getting flamed yourself, DON'T WRITE IN ALL CAPS; it's the e-mail equivalent to SCREAMING at people. Don't send a mass mailing advertisement (it's considered junk mail) to people you don't know; it can evoke a flaming response. Don't correct or reprimand someone in an e-mail and copy it to a number of others. That's the equivalent of yelling at someone in public and is always likely to get an instant rebuke—if only to save face.

Don't Continuously Read E-mail

This tip will help you manage your time and your writing. If you keep your e-mail on all day, you'll tend to check it constantly. Rather, plan to check it three or four times a day. That's sufficient to keep you up to date, but not constantly distracted. At a minimum, check it when you begin and end your day.

Re-read before Sending

Spell check and re-read every e-mail before ever hitting the send button. With important e-mails, read them out loud. The informality of e-mail can lull you into a false sense of security. Stay on guard to sending out something that could embarrass you later. And again, double check the "To" line.

Don't Send Chain E-mail

Just as chain letters are illegal in the U.S. mail (snail mail), chain e-mails are not permitted on virtually any network system and can get you removed from a system. Typically, such chain e-mails promise wealth, fame, or disaster if you don't pass them on. Trust me, delete them and don't pass them on, or you'll experience disaster sooner or later.

The good news is that e-mail is fast and effective. It has revolutionized our speed and efficiency at work. Today you can literally publish your ideas to hundreds or thousands of people with the keystroke of the send button on your e-mail system. You can reply any time and any place, and you can print out documents on demand. We've become much more effective because of e-mail.

The bad news is that we've also gotten much sloppier in what we write and to whom we send such slop. We consider e-mail somewhere between talking and writing. The problem is that most of us would not want a chance casual conversation to be reduced to paper and then later read back to us, or worse, read to many people who we don't know. E-mail has that latent, unattractive capacity.

E-mail is the killer application of the Internet. Use it wisely and enjoy it.

CHAPTER 12

FINAL WORDS

I've been a writer most of my life. It's been a long path with some successes and some failures. So I decided to make a list of my top ten writing tips.

Top Ten Tips

1. **Ask Two Questions**

   Whenever you start to write, ask two questions: Who is the audience? What's the purpose? These two simple questions will quickly focus your writing. You'll find that these are not always easy questions to answer. But once you do, the path is clearer and swifter—for you and the reader.

2. **Do Your Own Research**

   Accomplished lawyers warn new lawyers: "Do your own research." You must own the material to write about it, and there is no substitute for the learning process as a writer. In fact, "writing to learn" is as basic to a writer as breathing. You simply have to sit down, locate the relevant material, and master it.

3. **Be Brief**

   Don't take a page to say what you can in a paragraph or take a paragraph to say what can be said in a sentence. Make each word count.

4. Keep Your Style Simple and Direct

Well-educated people are often compelled to use complex, over-blown language. Don't. It's like trying to teach a pig to sing: He'll never learn how, and you'll only annoy the pig.

5. Make Writing Visual

Nobody likes to stare a long document in the face. In business, executives push such tomes to the side to be read later—much later. Make your documents attractive and easy to follow. Use subheads to break up the text. Provide the reader with "visual oases," natural places to stop and rest. Use transitions to move the reader in the direction you've headed. And use bullets, boldface, and italics to emphasize important information and create a visually appealing document.

6. Organize Your Writing

Organization has been mentioned repeatedly in this text because it's so important. Add order to writing. Here are some simple ways: If you're in a hurry, number the paragraphs. This alone makes it organized. Next, have an introduction, a body, and a conclusion. Finally, try to use a chronological approach: On Monday, then Tuesday, etc. Provide order, and readers will follow.

7. **Plan to Revise**

Revise your writing—period. Writing is always a work in progress until you send it off to the reader. Plan time to revise before you send writing to a critical audience. Rather than a sign of inexperience and weakness, revision shows maturity and strength. Often on deadline, it's difficult to find someone to work with you and your writing. So remember some hints already mentioned in this book: One way to look at your writing in a different way is to read it aloud. By listening to what you've written, you often hear where the errors and rough spots are. Another way is to reformat the text. If the original draft is single spaced, then double space it or place it in two columns. The visual difference helps give you distance and an objective eye.

8. **Ask Others to Review**

No matter how successful a writer you are, you have a host of personal biases and blind spots in your thinking—everyone does. To filter them out, ask others to read your writing before launching it on your readers. Look for people to review your writing, especially people who think differently than you. Getting a variety of opinions refines your writing and avoids embarrassing gaffes.

9. Just Write

Write if you want to improve. There's no substitute. If you want to get better at golf, tennis, swimming, or any skill, you have to do it. Nike's commercial hits it on the head: "Just Do It!"

10. Good Luck!

Consider this old saying: The harder I work, the luckier I get. This adage rings especially true in writing. So good luck and "Just Write It!"

ABOUT THE AUTHOR

Steve Gladis, Ph.D.
University of Virginia
The varied career of Steve Gladis includes academic, government, military, and civic service. Regarded as an educational entrepreneur, Dr. Gladis has a strategic goal for the University of Virginia in Northern Virginia: to link corporate needs with academic resources to better serve the community, the state, and the country. Dr. Gladis is a member of the University of Virginia's faculty and serves as both an associate dean in the School of Continuing and Professional Studies as well as the director of the University's Northern Virginia Center, which serves thousands of students through hundreds of courses and programs.

In his previous career, Dr. Gladis served as a Special Agent in the FBI. He was the chief of the Law Enforcement Communication Unit at the FBI Academy and was also the editor of the *FBI Law Enforcement Bulletin,* the most widely read law enforcement monthly journal in the world. Prior to the Academy, he served as the chief of speech writing for the director of the FBI and held a number of both headquarters and field-agent assignments around the country. He is also a former U.S. Marine Corps officer and a Vietnam veteran.

He has published numerous magazine and journal articles, as well as 11 books, including *WriteType: Personality Types and Writing Styles, Surviving the First Year of College, Effective Writing, Public Presentations,* and *Process Writing.* A regular lecturer and speaker, Dr. Gladis consults with corporations and organizations in the area of training and development.

A committed civic and academic leader, Dr. Gladis serves on the Executive Board of the Fairfax County Chamber of Commerce, is Chairman Emeritus of the Washington Math Science Technology Public Charter High School, and is a member of the University of Virginia's Faculty Senate.